AMAZING AGATES
LAKE SUPERIOR'S BANDED GEMSTONE

Written by
SCOTT F. WOLTER

Illustrated by
RICK KOLLATH

Kollath+Stensaas Publishing
394 Lake Avenue South, Suite 406
Duluth, MN 55802
(218) 727-1731

AMAZING AGATES *Lake Superior's Banded Gemstones*

©2010 by Mark Stensaas and Rick Kollath. All rights reserved. Except for short excerpts for review purposes, no part of this book may be reproduced or transmitted in any form by any means, electronic or mechanical, including photocopying, without permission in writing from the publisher.

Editorial Direction by Mark Sparky Stensaas
Illustration & Graphic Design by Rick Kollath, www.kollathdesign.com
Photography by Mark Sparky Stensaas
Special thanks to Stan Hendricksen for the original cover concept.

Printed in the United States of America by JS Print Group
10 9 8 7 6 5 4 3 2 1 First Edition

ISBN-13: 978-0-9792006-9-4

Dedication

*To my wife Janet, children Grant and Amanda,
and to all the loyal readers of my agate books
who have supported this agate obsession
for over 30 years.*

Contents

Introduction 1
Defining the "Laker" 2
How Do Lake Superior Agates Form? 3
Types of Lake Superior Agates 7
Agate Features 10
Equipment for Agate Picking 14
Where to Find Lake Superior Agates 15
What to Do with Lake Superior Agates 17
What's My Agate Worth? 19
Scott's Scale of Value 20
Conchoidal Fracturing and the Stone Age Hunter 21
Famous Lake Superior Agates 22
Other Types of Agates 27
This is NOT a Lake Superior Agate 30
Conclusion 31
Glossary 32
Index 34
Titles of Interest 36

AMAZING AGATES Lake Superior's Banded Gemstone

Introduction

If you purchased this book and are reading about Lake Superior agates for the first time you are in for a real treat. Some of you are reading while riding in a car pausing occasionally to view the incredible rock formations that make up the north-western coastline of the largest fresh water lake in the world: Lake Superior. It is within these rocks that the stunning gemstone you are about to explore was born over 1.1 billion years ago.

In November of 1986, I published the first edition of *The Lake Superior Agate*. I wrote the book as a personal healing project in the wake of my father's death, and as a way to share my knowledge and experience collecting what is arguably the most beautiful agate in the world. I enjoyed unexpected success to the point where the book is now into the second printing of the Fourth Edition.

When I was approached by Sparky and Rick about writing this book, I was hesitant at first. Why would I want to write another book about Lake Superior agates and compete with myself? After discussing the project with them along with my wife Janet, it became clear there was niche in the market that my previous books have missed. Sparky and Rick have enjoyed great success with their book, the *Rock Picker's Guide to Lake Superior's North Shore*, which has become a staple in bookstores and gift-shops along the shores of the big lake.

They believed a similar, more affordable book about agates would do well in the same market and I agreed. Unlike other agate books with black and white and/or color photographs, this guide is filled with beautiful full-color illustrations by Rick Kollath that bring these incredible gemstones to life in a way photographs cannot.

I am excited to share this information along with many wonderful stories about my favorite rock which in 1969 became the official State Gemstone of Minnesota. Before getting started, I do have to pass along a warning I have issued many times. The collecting "bug" for Lake Superior agates is highly infectious, and once it bites, its effects will last a lifetime!

Defining the "Laker"

For those just beginning, the first question you might ask is just what is a Lake Superior agate? Agates are comprised of quartz (SiO_2) which is non-crystalline or what geologists call "amorphous." The quartz actually occurs as tiny needles that line up in various ways such as clouds, groups or bands. It is the banded agates that are the most popular and occur in an infinite array of beautiful patterns.

When looking at the banding under magnification one can actually see the fibrous needles aligned perpendicular to the banding plane. Exactly how this occurs is a mystery that we'll discuss in a few chapters.

"Lakers" as many people call them, occur in virtually every color in the rainbow, but the most predominant colors for which they're recognized world-wide are the various shades of red and white.

Lake Superior agates were long known as the oldest agate in the world with their formation dating back over one billion years. While the final verdict isn't in, recent age-dating of other agates found around the world suggest the "oldest agate" title could change. Regardless, the remarkable age of Lakers —up to a billion years old—is mind-boggling.

Perhaps the most unique thing about Lake Superior agates is their glacial history. Huge mile-thick glaciers plucked out the agates from their bedrock host and carried them south and west of Lake Superior, spreading them across a vast hunting area. It was through the highly abrasive nature of that transport that the stones were broken open, exposing the myriad of banding colors and patterns.

Lake Superior agates can be found as small as sand-sized chips up to several pounds in size. The largest banded agate ever found in glacial drift weighed over 21 and a half pounds and was the size of a basketball! Most agates that collectors find range in size from a small marble to a golf ball. However, it is the bigger, brightly colored banded agates that people want. They're out there, but are very hard to find. The best way to help yourself find them is to learn as much as you can. That's where this book will help.

Lake Superior agates were pushed far from their origin during the last ice age. Some were even washed down the Mississippi to the Gulf of Mexico.

AMAZING AGATES   Lake Superior's Banded Gemstone

How do Lake Superior Agates Form?

The formation of the Lake Superior agate began 1.2 billion years ago during a geological time period called the Precambrian Era. During this time, the landmass that we now call North America began to split apart into two continents in what is called a *rifting event*. Internal forces within the earth produced upwelling lava which stretched the earth's crust. These forces split the crust open and poured out thousands of lava flows, some of which you see today from Lake Nipigon in Ontario, to Interstate State Park near Minneapolis-St. Paul. The rift zone itself was much bigger, extending eastward into what is now Michigan and southward into what is now Kansas.

Shortly after the rifting event started (15–22 million years in geological time is considered pretty short), it stopped.

The Midcontinent Rift

The Midcontinent Rift is a 1,240 mile long geological rift that runs through the center of North America. It formed about 1.1 billion years ago when the continent began to split apart. The rift failed, leaving behind thick layers of igneous (formed from lava) rock which are exposed along much of the north and south shores of Lake Superior. (After Chase & Gilmer, 1973)

1.1 Billion years ago, basalt lavas flowed over the surface

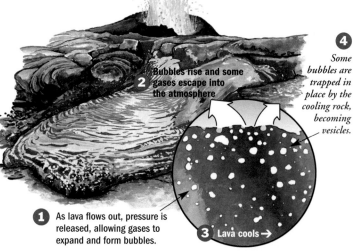

① As lava flows out, pressure is released, allowing gases to expand and form bubbles.
② Bubbles rise and some gases escape into the atmosphere
③ Lava cools →
④ Some bubbles are trapped in place by the cooling rock, becoming vesicles.

Highly compressed gases (mostly CO_2) in the lava expanded and formed millions of bubbles as the lava poured out. Some bubbles reached the surface and escaped; others were trapped as the liquid lava became solid rock.

Typical cross section of North Shore basalt lava flow

Note that vesicles form primarily at the top of a flow, but also along the bottom where steam is created when the lava rolls over water.

From vesicles to agates

① Over eons of time, groundwater percolated through cracks and fissures, filling the empty vesicles.

② The water contained silica and other trace minerals which were deposited in the cavities.

③ Eventually, many of the vesicles—filled with varying deposits of silica in layers—become agates.

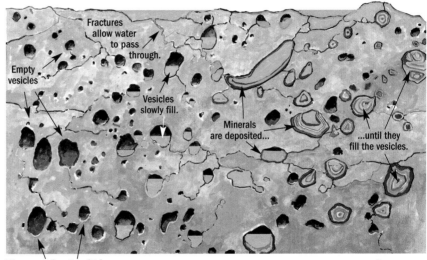

The gas pockets in which the agates formed tend to be small, about 1 cm in diameter.

By that time, enough lava flows had accumulated along the North Shore of Lake Superior that if piled up they would be over four miles thick!

As the lava flowed from fissures in the earth onto the surface, both water vapor and carbon dioxide gas rose toward the top of the flow. Most of these gasses reached the surface and escaped into the atmosphere. However, millions of bubbles became trapped as the lava slowly cooled. It was within these vacant gas pockets — or vesicles — that agates would eventually form.

Groundwater moving through the lava pile leached out minerals that migrated in solution through an intercon-

AMAZING AGATES LAKE SUPERIOR'S BANDED GEMSTONE

necting network of cracks and vesicles. Silica (SiO_2) is especially mobile in alkaline solutions and was deposited within the gas bubbles layer by layer. Some researchers believe agates form when the vesicle becomes filled with silica-rich gelatin-like solution, and then internally separates into the beautiful banding. However, there isn't a consensus amongst experts which is one of the mysteries and wonders of this amazing gemstone.

The next important period in the formation of Lake Superior agates began over two million years ago (essentially yesterday in geologic time) with the growth and expansion of massive continental sized glaciers. In North America, one of these large sheets of ice was centered over what is now Hudson Bay. At its maximum, the thickest ice was three miles thick and sent lobes of mile-thick ice into what is now the northern half of the United States.

The last ice sheet to scour the Great Lakes region—called the Wisconsin Glaciation—grew nearly three miles thick! It retreated 10,000 years ago.

Agates from ice

As the Superior Lobe scraped southwest through the basin of what later became Lake Superior, it wore down and ground up the ancient basalt lavas of the Midcontinent Rift.

How do Lake Superior Agates Form? **AMAZING AGATES** 6

The Superior Lobe followed the trough created by the over one-billion year-old rifting event. As the ice advanced it picked up the previously weathered out agates and plucked out and carried along gazillions more from the agate-filled lavas on the journey southward.

By roughly 10,000 years ago the glaciers melted away depositing massive quantities of mud, sand, gravel and boulders. Within these glacial deposits were millions of Lake Superior agates that were spread across a vast hunting ground that included parts of Minnesota, Wisconsin, Iowa, Nebraska and Kansas.

The myriad of beautiful banding patterns and color were revealed by abrasion during glacial transport.

The silica micro-crystalline agates were harder and thus more resistant than the basaltic bedrock, and often resisted crushing.

Agate

Volcanic rock and gravel comprised of basalt and rhyolite (pink).

Glaciology 101

Direction of flow

Base of ice sheet — ICE

Basalt bedrock with agate amygdules

Ground-up rock, pushed along for up to hundreds of miles.

Meltwater at the base of the glacier acts as lubrication, allowing the glacier to slide forward.

Glaciers are great landscape levelers. Here a small obstruction is being fractured and some of the agate amygdules embedded in the rock will be broken out.

It's the broken rock pushed along under the glacier — not the ice — that does most of the smashing and grinding of the bedrock.

Types of Lake Superior Agates

Invariably, when collectors begin to have success collecting banded Laker beauties they ask themselves, "How many different varieties of Lake Superior agate are there?" The answer is there are at least twelve main types: fortification, floater, paintstone, water-level, eye, tube, moss, sagenite, peeler, mosaic, blue or skip-an-atom and wave. Sometimes, if a collector is very lucky, they can find agates that combine two or more varieties.

Fortification Agate

The most common and easily recognized type of Lake Superior agate is the fortification agate. Because of glacial abrasion during transport the banding became exposed to produce an endless variety of beautiful patterns.

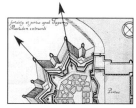

*The name **fortification** is used because the enclosing nature of the banding is similar to the walls of a fort.*

Floater

Another beautiful agate type is actually a sub-variety of fortification and is called a "floater." These specimens are produced when alternating layers of chalcedony bands appear to float between transparent layers of crystalline quartz.

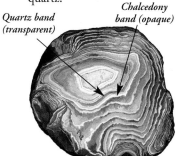

Quartz band (transparent)
Chalcedony band (opaque)

Agates with alternating chalcedony and transparent quartz bands ("floating" bands) are called floaters.

Paintstones

Painted agates or "paintstones" are fortification agates that are rich in various shades of opaque brown, red, orange or purple colors with white banding. The dense and rich colors that look as though they were painted on with a brush are produced by a high concentration of iron oxide pigment that was brought into the agate within the silica-rich solutions when it formed.

Paintstones typically have richly-colored, opaque banding.

Types of Lake Superior Agates **AMAZING AGATES** 8

Water-level Agate

Another variety that is actually a sub-class of fortification is the water-level, parallel banded or onyx fortification agate. "Water-levels" are likely produced by low fluid pressure within the vesicle during formation and can be comprised of only a few parallel bands or completely filling the entire agate.

Water-level agates show straight parallel banding.

Note that this rare agate was tilted at some point in its development through uplift or seismic forces.

Eye Agate

Arguably, the most popular variety is the eye agate. The perfectly round, usually banded shapes look like eyes staring back at you. Along with being the hardest variety to find in the field, their formation is the most mysterious and controversial. When cut into thin sections and reviewed under a polarized light microscope the chalcedony fibers of the "eye" radiate out from a single point to form a perfectly round shape. Why this happens is unknown. Some specimens may have up to fifty small eyes covering the surface.

Eye agates are highly prized. The bigger and more eyes, the better.

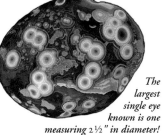

The largest single eye known is one measuring 2½" in diameter!

Tube Agate

The tube agate is a relatively rare and very beautiful variety. The tubes were formed by hair-thin projections of rod-like minerals usually aligned vertically and parallel to one other. As the chalcedony began to crystallize around these obstructions, the banding followed the contours of the projections, creating beautiful patterns. Tubes can also be encircled by massive clear quartz that, when polished, display beautifully. They vary in diameter and length from one millimeter to more than two inches in a few large specimens.

Tube agates are relatively rare and quite spectacular.

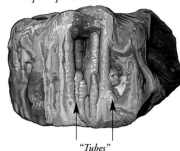

"Tubes"

Moss Agate

Perhaps the least appreciated variety is moss agate. At first glance these specimens don't look like much. But upon closer examination you can see dendritic or moss-like aggregates of material that either crystallized inside the vesicle before solutions entered the cavity, or were brought in along with the solutions. Under magnification there are a myriad of tiny banding patterns between the moss-like projections that can often resemble figures and landscapes.

This cut and polished moss agate was found in an Iron Range gravel pit and originally weighed 17 pounds.

AMAZING AGATES TYPES OF LAKE SUPERIOR AGATES

Sagenite Agate
Sagenite is a special variety that contains needle-shaped mineral inclusions that typically form radiating "bushes" or "sprays." The jet black to golden copper colored needles typically originate from a central point with well-defined, straight crystals of rutile, goethite or both.

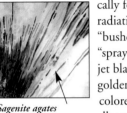
Sagenite agates typically display radiating, needle-shaped formations and range greatly in color.

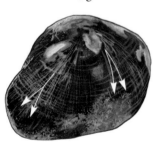

Peeler Agate
A common feature that is easy to spot when agate hunting is called "peeled" texture. It is called this because it appears as though the bands were peeled off like a banana skin. This tell-tale sign of smooth exposed band planes, along which the agate has fractured or broken through glacial abrasion and freeze-thaw cycling, are sometimes readily visible in dirty, poorly exposed rocks.

Exposed banding

This "peeler" has exposed banding that resembles sedimentary layers of rock exposed by erosion.

Mosaic Agate
Mosaic agate is a sub-class of water-level agate that has a polygonal pattern on the bottom side. These agates were likely produced by drying shrinkage cracking of the first horizontal band deposited in the gas pocket during formation.

The broken glass appearance of mosaic agates resembles a tiled mosaic or a mud puddle crack pattern after drying in the sun.

Wave Agate
Very rare. Has gently undulating red and white banding snaking around the husk of the stone.

Blue Agate/Skip-an-atom
A new variety of Lake Superior agate burst onto the scene in the late 1990s, called "Blue" or "Skip-an-atom" agates. These agates have a unique and distinct light bluish-gray color with fibrous crystalline quartz that is often bright white in color.

"Blue" or "Skip-an-Atom" agate

Wave Agate

Agate Features

As wonderful as agates are when you hold them in your hands, there is another level of complexity and beauty when agates are explored under magnification. Use a hand lens or microscope! There are many incredible structures to be seen if you simply take a closer look.

Banding

Most agates upon close examination contain an easily recognized structure called similar banding sequences. The typical sequence of the classic "red and white" is usually comprised of fibrous chalcedony needles aligned perpendicular to the banding plane in the white band. Next is a relatively clear or transparent zone that often contains copper and sulfide minerals. Last is an iron-oxide-rich red or orange layer that is usually quite thin relative to the white and clear layers.

Copper

Copper can sometimes be quite prevalent usually occurring in small aggregates that align perpendicular to the banding plane between the chalcedony fibers. In the Upper Peninsula of Michigan near the town of Calumet, there is a mine that produces crushed basalt that contain small agates with banding that is solid copper. Extremely rare and valuable specimens with copper eyes have been collected.

Copper found in agates can take many forms from bands (below) to minute grains, to leaf-like crystal structures (above).

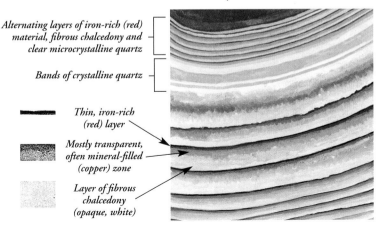

Alternating layers of iron-rich (red) material, fibrous chalcedony and clear microcrystalline quartz

Bands of crystalline quartz

Thin, iron-rich (red) layer

Mostly transparent, often mineral-filled (copper) zone

Layer of fibrous chalcedony (opaque, white)

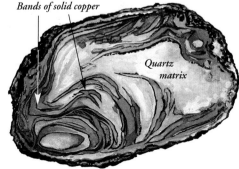

Bands of solid copper

Quartz matrix

Iron oxide

Perhaps the most common microscopic feature in agates is iron oxide particles that were carried in with the silica-rich solutions when the agate formed. Most often the particles are brown or red in color and typically occur as fine disseminated particles along the banding. The predominantly red, orange and yellow colors are produced on the surface of the agate by weathering when the iron oxide particles within the chalcedony bands are exposed.

Iron oxide particles often appear in a thick layer...

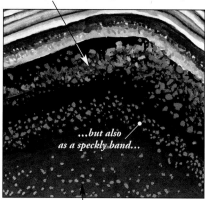

...but also as a speckly band...

...or as a field of red–orange–yellow specks.

Shadow

Many agates have well developed, very fine banding made of alternating opaque white and clear layers. When these bands are aligned parallel to the viewer the eye perceives depth in the agate. This eye-catching feature is called *shadow effect* and is relatively common in Lake Superior agates. Because of the oxidation colors caused by weathering on the surface of agates, shadow banding is often obscured. Many otherwise marginal agates reveal beautiful shadow when polished.

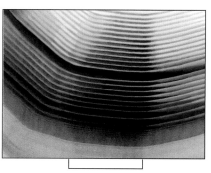

Shadow effect

Pinching Banding

Another common feature of fortification agates is pinched banding. Banding thicknesses within a single agate can vary dramatically even within an individual band. If you trace a single band around the pattern of the stone it will almost always get thin in places and often seems to disappear. Why this happens is not something geologists agree on, but many believe it is related to the velocity and viscosity of the silica-rich solution as it moves when the agate forms.

Wide banding

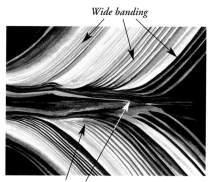

Pinched banding

The banding appears to disappear when it pinches near fill holes. The banding is still present but is too thin to be seen with the naked eye.

Fill Holes

Commonly seen are entrance channels or fill holes that appear to be where the majority of solution flowed into, and out of, the agate during its formation. Along these channels the banding typically deforms toward the exterior of the agate as if silica gel was being pulled from the interior of the agate.

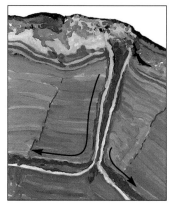

This fill hole allowed the agate to continue forming banding at the center.

Crystal Impressions

Crystal impressions are formed when other minerals form inside the empty vesicle before agate formation begins. The most common minerals that crystallize first are calcite, aragonite, barite, rutile and zeolites. Once these crystals form, silica-rich solutions enter the vesicle producing an agate with banding patterns around the previously formed minerals. After the agate was freed from its host rock and transported by glaciers, the minerals that formed the crystal dissolved, leaving a mold or cast on the surface (or *husk*) of the agate.

Cast of missing crystal

Water-washed

As the name implies, these well-rounded specimens were likely rolled around in a river or the surf of an ancient beach until highly polished and smooth.

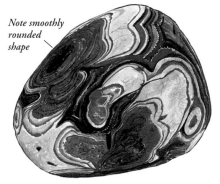

Note smoothly rounded shape

This beautiful water-washed specimen was used by my wife, Janet, as a "focal point" during the birth of our son Grant. We affectionately call this agate "Baby Heart."

Ruin

Good examples of ruin agates are extremely rare. They were formed while the agate was still embedded within the host lava flow. Earthquakes associated with volcanism caused faulting and cracking of the flows and sometimes also broke the agates within them. Most specimens of ruin agates exhibit displacement of banding that can be quite dramatic and likely related to the intensity of the tremors. The cracks running through the agates were subsequently cemented back together by later solutions of chalcedony. The name ruin agate was given because the otherwise intact agate was "ruined" by the fractures. On the contrary, ruin agates are some of the most interesting and rare of all agate types!

Displacement banding: offset bands, like a mini fault line, distinguish this fractured and recemented agate.

Breccia

Breccia agate is formed in the same manner as ruin agate, only the earthquakes caused the rock to break up into pieces. Subsequent silica-rich solutions cemented the sharp, angular fragments together into a solid mass.

The word breccia *comes from the same Germanic root as the English* break. *It's obviously the right word choice for these kinds of cracked and shattered agates.*

Candy Striper

Candy stripers are highly prized and are comprised of strikingly bold, alternating red and white banding. Lake Superior agates are known throughout the world for this classic variety.

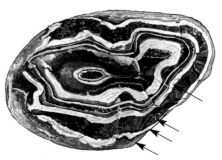

Note alternating red and white "candy stripe" layers

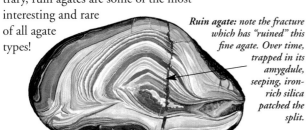

Ruin agate: note the fracture which has "ruined" this fine agate. Over time, trapped in its amygdule, seeping, iron-rich silica patched the split.

Equipment for Agate Picking

The equipment needed for collecting agates depends on how serious a collector wants to get. In some cases, all one simply needs is a pair of shoes that can get dirty, a sharp eye and place to pick. However, for those who are more serious and are planning to pick for several hours or a full day, then additional items will make for a more enjoyable and fruitful day.

Magnifying glasses typically have a power of about 2x. A loupe (or "hand lens") is a fairly powerful magnifying glass— typically 10x. It's also a nifty gadget for enjoying the banding of tiny agate specimens. Don't dismiss the small and broken agates as they may show the same features as agates many times larger.

Since many agates have wonderful features which can only be appreciated under magnification, bring at least a magnifying glass.

Besides finding a suitable area to collect, the first priority is always the weather. Check the forecast and then wear the appropriate clothing. Dressing in layers makes good sense and always bring a rain jacket. Most agate collectors hope for rain to wash off exposed gravel making it easier to find the coveted gems. Hiking boots are excellent for long treks across various types of agate-collecting terrain. Dry socks are the golden rule and be sure to bring along an extra dry pair in case your feet unexpectedly get soaked.

On sunny days a hat is needed to prevent a sunburned face or to keep your head dry if the skies open up with rain. Other items you might want to bring include sun-block, a compass, a hand-lens or 10x loupe, a trowel or hand-rake for sifting through gravel piles and a bucket or pack to carry the finds. These days nearly everyone carries a cell phone: that might help if you get lost, but in most cases, you'll use the phone to immediately call someone if you happen to find a lunker laker!

Don't forget to pack a lunch or some snacks to replenish your energy and enough water to keep your throat from getting parched. Of course you'll also need a backpack to carry your food and spare gear. Hopefully, you'll also need room in your pack to carry home the gorgeous agates you worked so hard to find.

A hand rake makes a great gravel-sifting tool. The longer the handle, the better!

You'll be lugging rock so bring a sturdy pack.

Where to Find Lake Superior Agates

The most fulfilling part of agate collecting is the challenge and joy of finding them in the field. There is nothing quite like the euphoria of discovering a beautifully banded specimen yourself. The distribution of Lake Superior agates is directly the result of glacial ice that deposited the gemstones primarily into moraines and outwash associated with meltwater streams. Agates are commonly found in gravel layers or beds that are exposed at the surface.

Please get permission from the land owner before entering their property. An "abandoned" gravel pit is probably less abandoned than you think!

Gravel Pits

By far the best place to collect Lakers is where these sand and gravel layers are exposed from mining the material in an active gravel pit. As a pit is worked, layers of sand and gravel are exposed. As the pit deepens, rocks fall from the steep walls and accumulate at the base of the slope. This concentrated gravel and rock is the most desirable place to "pick" or look for agates and is called "the drop."

As exciting and productive as theses areas are, caution is necessary. The recently exposed walls are very unstable and often rapidly cascade down without warning and can bury a person in seconds. Stay off the high banks and keep clear of steep, recently worked areas when looking for agates in gravel pits. Remember that gravel pits are private property, and the only acceptable way to hunt for agates is to have written permission from the owner.

Construction Sites

Construction sites are another good place to collect because bulldozers and backhoes expose gravel beds during the shaping of the landscape. These sites can be excellent for collecting especially after a cleaning rain washes the stones at the surface.

Urban and Suburban

Decorative landscaping stone produced in gravel pits is commonly used around residential homes and other buildings, parking lots, trees, bushes and elsewhere. Many people take this decorative landscape rock for granted not realizing that many, fine, unseen gems lie waiting to be discovered.

Lake Superior Beaches

Another good place to search for agates is along gravel beaches and shorelines of Lake Superior and other lakes. Many people become discouraged picking along the shores of the "Big Lake" due to decades of previous collecting. However, successful picking can be had after high winds or a storm produce large waves that stir up the rocks in the shallow water and push agates up onto the beach.

Where to Find Lake Superior Agates

Streambeds

The vast Lake Superior agate collecting area has thousands of rivers and streams cutting through the countryside and is another great place to search for agates. The best time to pick is after a heavy rain or the spring thaw. Rivers and streams will erode the glacially derived material and periodically free agates to be discovered along the freshly washed banks.

Farm Fields

Farm fields, especially in the spring before the crops come up are another great place to collect agates. Hundreds of thousands of acres of farmland are littered with agates that are continually brought to the surface by freeze-thaw cycling in the ground. Some of the largest and most beautiful Lake Superior agates ever found came from farm fields and the rock piles made by farmers clearing their fields. Here again, one must ask permission to collect anything on private property. In most cases, farmers are more than happy to let people collect rocks from their fields so they don't have to!

Gravel Roads

Another great place to find agates, albeit mostly smaller ones, is while walking along dirt and gravel roads after eating lunch or dinner at the cabin or while vacationing at a lodge. "Agate country" has thousands of miles of unpaved roads and trails filled with gravel that is peppered with Lake Superior agates. The best time to wander the roads and trails is just after, or better yet, during a rain. Even in the pouring rain it's fun to throw on a jacket and scan the freshly washed sand and gravel roads for agates.

LAKE SUPERIOR BEACH RECYCLING 101

Gravel beaches, especially along the north and south shores of Lake Superior, can be productive even though they get repeatedly picked over for agates. Powerful waves generated by frequent storms act like a giant rock tumbler reworking the beach gravel, exposing new, usually well-rounded agates for collectors to find.

What to Do with Lake Superior Agates

Many people wonder what can be done with the agates that they collect. For the finest specimens with high quality banding exposed by natural glacial action should not be cut or polished. However, some agates need a little help. Lapidary—cutting and polishing rock—can sometimes be used to enhance your finds. Here are a few ideas.

Oil Treatment
The safest thing to do with your prized gems is to clean them thoroughly with soap and hot water and then enhance their beauty through a process known as *treating*. The hot water heats the specimen and slightly opens the tiny surface micro-cracks. The next step is to dry the agate with a towel and then immerse it in "baby" or mineral oil while still warm.

This allows the oil to seep into and fill the minute cracks, letting light reflect off the surface. Once the oil has seeped in, been wiped off and allowed to dry out, all the beautiful banding patterns and colors will be rich and full.

While I am personally partial to oil-treated rough agates, there are definitely some Lakers that "need a little help." Lapidary is the art of cutting and polishing rocks, and some Lake Superior agates can take on added beauty and value when polished or cut. Agates, like diamonds, must be individually studied to determine how to bring out their highest beauty and best optical properties.

Tumbling
The most common lapidary technique used on agates is tumbling. A tumbler consists of a belt-driven hollow drum containing polishing grit. The motor slowly turns the drum filled with agates that tumble in the grit. Over a period of days or weeks, the grit slowly abrades the agates exposing their hidden beauty. Newer tumblers have vibrators that reduce the time necessary to smooth-finish the gemstones. Tumbling is usually recommended for smaller agates that range in size from ½" to 1 ½". Most rock shops and hobby stores carry rock tumblers that range in price from $100 to $1000.

Lapidary tumbler

Cabbing
Cabbing is the making of cabochons or "cabs" from slabs cut from agates. Ovals or other shapes are traced out and then cut from

the slabs and then polished. Finished cabs can be used to create rings, belt buckles, necklaces, earrings, and other types of jewelry. Cabbing is a very delicate art and requires skill and experience to do well.

Slab Cutting

A technique used on larger agates is slab cutting. Cutting slabs is done using diamond-studded circular saws with blades ranging from 4" to 20" in diameter. Larger, poorly exposed rough specimens are suited for slabs that can be made into ash trays, wind-chimes, tabletops and cabochons.

Agate cabochon

Face Polishing

Another popular lapidary method typically used on medium to larger agates is face polishing. Creating a "face" on an agate involves cutting a portion of the stone, or grinding down a desired area and then polishing the surface. Often, an exposed surface will be heavily oxidized masking much of its internal beauty. These kinds of agates will be greatly enhanced by face polishing. These specimens are interesting because most of the agate is left in its natural state with only one side having a handsomely polished face.

Marbles

An interesting but relatively rare lapidary technique used on agates is the production of marbles and spheres. These amazingly beautiful gemstones are perfectly round and require a great deal of skill to produce.

A selected specimen is first cut into a cube. The corners are then cut off and the specimen is ground into a rough sphere. The final step is to take the "near-sphere" and place it into a sphere polishing machine comprised of three polishing heads that complete the process.

Patience and care are required before any lapidary procedure is started. Because of the violent nature and repeated freeze-thaw cycles associated with glacial transport, fractures are all too common in Lakers. Always carefully study an agate for fractures and banding patterns before deciding what to do, for once an agate is cut or tumbled, you can never go back.

Agate marbles

Sources for
LAPIDARY SERVICES

Don't want to do the work yourself? There are several rocks shops that will do custom polishing work.

Agate City Rock Shop *Two Harbors, MN*
Specializing in Lake Superior agates, books, jewelry, gifts and lapidary services.
(218) 834-2304
www.agatecity.com

Beautiful Agates *Minneapolis, MN*
Specializing in Lake Superior and world agates, books and lapidary services.
(612) 929-5779
www.beautifulagates.com

Beaver Bay Agate Shop *Beaver Bay, MN.*
Specializing in Lake Superior agate, Thomsonite, Isle Royale Greenstone, lapidary equipment, jewelry, carvings, fossils & gifts. (218) 226-4847
www.beaverbayagateshop.com

Grand Marais Rock Shop
Grand Marais, MN
Specializing in Thomsonite, agates, amethyst and lapidary services.
(218) 387-9291

Rocks & Things *Princeton, MN*
Retail rock shop and home of Minnesota Lapidary Supply. (888) 612-0405
www.rocksandthings.com

What's My Agate Worth?

One of the most commonly asked questions by people who have found an agate and are thinking of selling it is: How much is my agate worth?

Although there isn't a general industry standard for valuing agates, the price starts at about 50 cents a pound for unsorted, tumbling-grade material. From there, prices typically range from $1 to $5 per pound depending on the average size and quality. After that, assigning values to agates gets a little more complicated with prices ranging from $10 a pound to as much as $1,000 a pound—which is retail pricing—for the absolute best, highest-quality, largest specimens.

One way to establish a value for larger, high-quality agates is to assign a ranking, from one to ten for various aspects including size, shape, color, banding, composition and contrast and the amount of undesirable crystalline quartz. The same criteria can be applied to other agate types like tube, eye, moss, water-level, sagenite, etc.

This ranking system has a good amount of subjectivity involved by the appraisers, but in almost all cases the bigger, higher quality agates will end up with higher scores no matter who does the evaluation. What I'd like to do is try to introduce a system that quantifies various aspects of Lake Superior agates to help appraisers determine a fair market value.

Size

For a long time, size was the primary factor in determining an agate's value. Size is very important, and after about one pound, the value goes up exponentially, due to rarity as size or weight increases. The numerical scale on the following page reflects the relative rarity (and value) according to size.

In theory, once an agate has been rated according to the six categories, the assigned numerical total will help in determining a value to the agate.

In the end, this ranking serves as only a guide, not the "gospel." The value of an agate is in the eye of the beholder, and, of course, the seller. Most often it boils down to the age-old axiom in sales that says an agate is worth what someone is willing to pay for it!

Scott's Scale of Value

Weight	Rating
Less than 0.5 lb.	1
0.5 to 1.0 lb.	2
1.0 to 2.0 lbs.	3
2.0 to 3.0 lbs.	4-5
3.0 to 5.0 lbs.	6-7
Greater than 5.0 lbs.	8-10

Shape	Rating
Highly angular with sharp edges	1-2
Irregular with rough edges	3-4
Irregular with smooth edges	5-6
Oval or flat with rough surfaces	7
Oval with smooth surfaces	8-9
Oval or round water-washed	10

Color	Rating
Light to dark gray	1-2
Yellow (limonite staining)	3
Rusty red & white	4-5
Dull red and white	6-7
Multi-colored "paint"	8
Bright red and white	9-10

Banding*	Rating
Undifferentiated	1-2
Weak or poorly defined	3-4
Fine to thick with low contrast	5-6
Fine to thick with medium contrast	7-8
Fine to thick with high contrast	9-10

also quality of eye, tube, etc.

Composition & Contrast	Rating
Broken or badly chipped	1-2
Mostly peeled and rough	3-4
Mostly peeled and smooth	5-6
Open, rolling pattern with husk	7-8
Complete fortification pattern with husk	9-10

Quartz	Rating
More than 75%	1-2
50% to 75%	3
25% to 50%	4-5
0 to 25% (solid plug)	6-7
0 to 25% (floater)	8-9
0% (solid agate, no quartz)	10

For Example...

When all aspects were rated, this 1.6-pound specimen received a score of 38 out of a possible 60 points.

Weight	3
Shape	9
Color	8
Quartz	5
Composition & Contrast	7
Banding	9
Total:	**41**

Shape {9}
Color {8}
Quartz {5}
Composition & Contrast {7}
Banding {9}

AMAZING AGATES — Lake Superior's Banded Gemstone
CONCHOIDAL FRACTURING *and the* STONE AGE HUNTER

Stone Age humans knew their stones. They had to. Survival depended on it. Only certain rocks could be made into sharp tools such as axes, hide scrapers, spear heads and arrowheads.

There is a trait specific to rocks composed of microcrystalline quartz called "conchoidal fracturing." Most minerals inherently have weakness planes in their atomic structure. When enough force is applied they break along these planes. This is called cleavage. Agates and the other microcrystalline quartzes lack these planes of weakness, having equal bonds of strength throughout, and instead fracture when broken.

How do you recognize conchoidal fractures? These rocks chip or break into concave half-moon or clam-shaped fractures. The clam analogy can be taken one step further; each fracture shows concentric ribs like the ribs on clams or fan-shaped seashells. In fact, the word "conchoid" comes from the Greek word meaning sea shell.

Ever seen a windshield that's been hit by a BB or small rock? The strike causes a shock wave that expands resulting in a cone-shaped break. This is essentially what conchoidal fracturing is, and it is what happens when microcrystalline rocks are hit with enough force.

Knapping (making tools from stone) was a critical skill for all Stone Age hunters. You needed to make a razor sharp arrowhead or spear head to pierce the hide of the mastodon, mammoth or moose. But how do you turn a piece of agate, flint or jasper into a razor-edged killing machine? Once a suitable stone had been found, the maker would shape the tool with a hammerstone. Once the tool has been roughly formed, the knapper would use a technique called pressure flaking to thin the tool and hone the edges to razor sharpness. In essence the knapper was making the edge from dozens of tiny, controlled conchoidal fractures. Working with the tool on his thigh or palm, he would apply pressure at an angle to flake off small chips.

Sources for such stones were likely well-known and highly revered—possibly even protected. After all, each band of hunters had to make hundreds of these projectile points each year.

In the area where Lake Superior agates are found, humans were not around until the last of the glaciers receded, about 12,000 to 10,000 years ago. At this time Minnesota was an open plain of grasses, spruce parklands, and large meltwater lakes and rivers where now-extinct animals roamed. Some Paleo Indian spear heads have been found here which were very symmetrical, requiring great knapping skill to create. These projectile points were made in Minnesota until contact with fur traders in the 1700s.

And just so you don't think we are quite out of the "Stone Age" yet, consider this; Obsidian scalpels are sometimes used in eye and cardiac surgery. They are many times sharper than the finest surgical steel scalpels with the edge being only 3 nanometers wide.

Knapping: Using the tip of a deer antler to fine-tune the edge of an agate arrowhead.

Conchoidal fractures: Note the concave half-moon or clam-shaped fractures and their concentric rings on this piece of obsidian.

A lovely knapped arrowhead made from a Lake Superior Agate

Rock types that exhibit conchoidal fracturing:
Crystalline Quartz • Agate • Chert • Carnelian • Chalcedony • Flint • Jasper • Volcanic Glasses • Obsidian • Silicified Limestone

Famous Lake Superior Agates

Counsell
Ernie Counsell was riding his tractor through a field on his farm near New Haven, Iowa in 1958 when he noticed something strange. Two young boys were taking turns trying to carry a large rock across the field. When he stopped to investigate, the boys said they found the rock along a creek running through the Counsell farm. Ernie recognized the rock as a huge Laker and told the boys it belonged to him. It turned out to be the largest Lake Superior agate ever found in glacial drift and weighs a whopping 21.62 pounds!

Ham
Sometime prior to 1921, Nina Cox was strolling through a pasture with her husband near Ellsworth, Wisconsin when she stumbled on this 10.25-pound monster. Until her death in 1981, at the age 93, Nina believed that what she had found was a petrified ham. Many collectors believe this fantastic specimen is the finest quality big Laker ever.

Batman
Twelve year-old Cody Post found this 4.74-pounder with his family in a gravel pit near Pierz, Minnesota on August 19, 2006. The bold fortification pattern reminds me of the "Batman" signal from the famous comic book character's television series.

Counsell

The "petrified" ham

Batman

AMAZING AGATES   Lake Superior's Banded Gemstone

Minnehaha Falls

Minnehaha Falls
My wife Janet and I were ecstatic the night we drove home after purchasing the famous 7.39-pound Minnehaha Falls agate that was found below the falls in South Minneapolis in 1932. It was in the car that we first noticed the shallow, carefully pecked groove that encircled three-quarters of the stone. Several experts said it was a sacred object fashioned by Native Americans in the past and likely was intentionally left below the waterfall after a special ceremony.

Berghius
In 1976, Lyle Berghius found this exquisite 6.38-pound gem while picking with his 10-year old daughter. The agate was used as part of a stone ring circling a campfire in a gravel pit near Milaca, Minnesota.

Berghius

Peanut Butter & Jelly
Peter Kreminski found this 2.24-pound stone in beach gravel along the shore of Lake Superior just north of Two Harbors, Minnesota in 1974. Unable to find a buyer for the nodule that had essentially no banding showing, he decided to cut it. Its amazing color scheme and fracture-free polished face makes it one of the most rare and popular Lake Superior agates ever.

Engagement
In 1986 when Janet and I were engaged to be married, she wanted to get me a unique engagement gift. She quietly worked a deal with our long-time agate mentor and friend, George Flaim. During a trip to Duluth that summer, she gave me the incredible gift while we were having a picnic on the North Shore of Lake Superior. This amazing 5.75-pound beauty was found by a boy scout, walking along the beach at Island Lake, Minnesota, sometime around 1930.

Peanut Butter & Jelly

Engagement

Rolling Stone

Often an agate banding pattern will look similar to a well known object. People always smile and laugh after seeing this 0.19-pound agate that looks exactly like the tongue logo of the legendary rock band The Rolling Stones.

Rolling Stone

Reef Agates

In July of 2007 I hosted an agate hunt for the *Cash & Treasures* show and took the host, Kirsten Gum, scuba diving on an underwater gravel reef in Lake Superior. The weather was horrible that day and shortly after jumping into the water we all climbed back out because of the rough water without collecting even one agate. Later that summer, my friend George Twardzik dove on that same spot and pulled out a beautiful 2.16-pond water-washed paintstone. Not to be outdone, the following May, while diving on the same gravel bar with my buddies George and Jake Anderson, I pulled out an exquisite 1.75-pounder. It was a glassy blue color with fantastic shadow banding in the face. This agate just might be the best Lake Superior agate I've ever found!

Scott's Reef Agate

George's Reef Agate

Red & White

Perhaps the finest example ever of a large Lake Superior agate with classic red and white colors is a 6.16-pounder found near Randall, Minnesota, around 1977.

Red and White

Remember Me?

The biggest game of "cat and mouse" I ever had pursing agates was one with young boy named Jon Morell. The eight-year-old first appeared with his 2.22-pound agate at the 1998 Moose Lake Agate Days Gem & Mineral show. After turning down my offer of enough money for him to buy a new bicycle, he disappeared into the crowd without my getting his phone number. I had taken a picture of him and published it in my latest agate book asking for another

chance. I didn't hear a word until 2003, when I received a picture in the mail with no return address. There he was holding the agate with an impish grin. I found a note on the back with the words, "Remember me?" I heard nothing again until June of 2006, when I received an email that read, "I'm ready to sell." After eight long years we finally met again. After consummating our deal we took one last picture of the now sixteen-year-old young man holding his agate for the last time with that same impish grin.

Doroff

Doroff
Jim Doroff was walking in the rain along a dirt road in the Foothills State Forest hunting when he first spotted the large banded stone in 1985. "The road had recently been straightened with a dozer and I walked past the agate intending to pick it up on the way back. I almost forgot about it, but picked it up and set it outside our hunting shack. I almost forgot it again when we left for home." I was elated that Jim remembered to pick up the 5.67-pounder which I bought on Valentine's Day 2010. Jim's find is one of the best big Lakers of all time.

Howling Wolf
I received a call from a man named Dave Kirkorkian who was interested in selling a 4.02-pound agate that he found while walking his dog in a gravel pit near Scanlon, Minnesota in 1994. My initial skepticism vanished when I saw the distinctive pattern in the agate he called the "Howling Wolf."

Howling Wolf

Howard
Matt Howard was sixteen years old when he found this 4.58-pound stunner while walking into a gravel pit with a friend near Saginaw, Minnesota in 2001. Matt wanted enough money for a down payment on a car and we both were

Remember Me?

Famous Lake Superior Agates

Howard

happy after we made the deal. I always bring the agate to Moose Lake Agate Days knowing that Matt will usually show up to again see the amazing treasure that might be the finest big Lake Superior agate I've ever seen.

Home Run

Nobody was interested in buying this two-and-a-half pound nodule with virtually no banding showing at all. I decided to gamble and bought the agate from my friend John Harris and agonized for weeks whether to cut it or not. Finally I decided to cut a slice off and hit a "Home Run" when the stunning inner beauty was revealed.

Home Run

Perfect Pounder

The striking red and white, concentric banding in this 0.96-pound classic beauty is exposed on both ends of the stone. One face is rough and the other is polished and because of its rare, eye-catching beauty it was nick-named the "Perfect Pounder." Bobby Schussler found the agate in a gravel pit in Lakeland, Minnesota in July of 1979.

Perfect Pounder:
Rough face (inset, above),
polished face (right)

Brush Fire

Jim Rodeski reportedly found this 2.69-pounder while helping a farmer dig fence post holes near Swanville, Minnesota in 2004. Jimmy walked around the Moose Lake show that summer showing off his new prize like a proud parent with a newborn child. The wild pattern and brilliant red-and-white colors of this stone inspired the name "Brush Fire."

Brush Fire

Other Types of Agates

While this only slightly biased author believes the finest agate in the world formed in the lavas along the shores of Lake Superior, I must admit there are several wonderful agates found in other states and other countries. There are dozens of other agate types, but here are a select few that should be mentioned.

Laguna Agate
Many people think Laguna agates from Mexico are the highest quality agate in the world, and they might be right. Formed primarily within the gas pockets in rhyolite lava flows, these nodules are mined from weathering flows and need to be cut open and polished to reveal the beauty inside. Laguna agates exhibit virtually every color in the rainbow with sharp, well-defined banding patterns that usually contain exquisite shadow effect.

Unlike most Lakers, Laguna agates must be cut and polished to reveal their inner beauty.

Fairburn Agate
In recent years, "Fairburns" have skyrocketed in value due primarily to their rarity and beauty. Fairburn agate formed inside solution cavities within sedimentary rock found outcropping at the surface around the Black Hills of South Dakota. These agates have a unique array of "earthy" colors exposed through eons of abrasion from transport by rivers and streams draining in all directions.

While Lakers are famous for being mostly red, orange and white, Fairburns typically exhibit "earthy" colors such as blue, yellow, gray, orange and black.

Montana Moss Agate
Usually light blue, white or translucent in the rough, Montana moss agate formed by silica replacement of wood buried beneath volcanic ash from eruptions millions of years ago. While not as colorful or impressive in their rough state, cut and polished specimens often contain black dendrites that weave through the banding producing stunning shapes and patterns.

Laguna

Fairburn

Rough

Polished

Montana agates are predominantly translucent gray with black and white banding in the rough. They can be quite striking when cut and polished.

Dryhead Agate

Dryheads are also found in south-central Montana and are a close cousin to Fairburns as they also formed within cavities of fine-grained sedimentary rock or volcanic ash. Like Lagunas, these agates have to be mined on site and then cut in half to expose beautiful pastel-colored banding patterns. Dryheads typically contain numerous fractures, but flawless examples do occur and can be spectacular.

After Dryheads are cut they usually exhibit orange and red colors, sometimes with striking white banding.

Botswana Agate

In the African country of Botswana, a beautiful banded volcanic agate is found that exhibits a unique array of blue, white, gray, lavender and pink banding. In addition to a myriad of beautiful banding patterns, some of the best examples of shadow banding result when specimens are cut and polished.

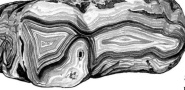

This relatively flat specimen with banding exposed on all sides is called a "Sandwich" agate.

Brazilian Agate

By far the most abundant and largest agates in the world are mined in South America. For over one hundred years, thousands of barrels of agates have been shipped out of Brazil to be cut and polished by professional and amateur lapidary enthusiasts. While "Brazils" have predominantly blue, gray and white colors, spectacular specimens have been cut with incredible colors and patterns. Most range in size from 1 to 2 pounds, but a few rare specimens reach up to hundreds of pounds in size!

Brazil agates have the greatest frequency of both water-level and fortification banding.

Australian Agate

Australian agates are primarily found on the northeast coast of the continent in relatively remote areas. These agates also form in volcanic rocks and typically have rich red, pink, yellow, green and purple colors. Cut and polished specimens usually are smaller in size ranging from a couple ounces up to a pound or so.

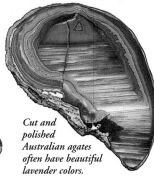

Cut and polished Australian agates often have beautiful lavender colors.

Chinese

A relative newcomer on the agate collecting scene are the spectacular water-washed and beautifully banded specimens imported from China. These agates are reportedly collected by farmers in gravel beds along the Yangtze River. Usually ranging from marble size up to a pound, these agates have been highly polished by extensive river transport and are often confused with high quality water-washed Lake Superior agates.

Thunder Bay Agate

We'd be remiss if we didn't mention the kissing cousin to Lakers found just over the border in Canada at Thunder Bay, Ontario. Within the local volcanic rocks is a vein type of agate with predominantly red, orange, and white colors. It occurs commonly as tubes with fortification patterns and has been mined out in chunks weighing up to tens of pounds in size.

The beautifully smooth myriad of banding patterns and colors make Chinese agates the finest water-washed agates in the world!

Thunder Bay agates form orange and white banding inside irregular-shaped voids and veins in volcanic rock.

This is NOT a Lake Superior Agate

In the early days of my collecting, I went off on more than one wild goose chase for an agate that turned out to be chert, jasper or some other variety of quartz-rich rock. I still get many calls from people who think they've found a monster Laker. § An attorney once wrote me insisting he knew the whereabouts of the biggest Lake Superior agate in the world. He said it was over four feet across, had excellent banding, and estimated its weight at several tons. When we met for lunch and he showed me the pictures I regretfully informed him that his record agate was actually a silicified limestone. Although it was a very pretty rock, it didn't yield the big pay-day he was hoping for.

It's easy for inexperienced collectors to misidentify agates since the dominant composition of many other rocks besides Lakers is fine-grained quartz. The physical properties of many varieties of quartz-rich rocks are very similar to agate such as a waxy luster and glassy appearance. The majority of rocks commonly mistaken for Lake Superior agate generally fall into the following categories:

Iron Formation
People commonly mistake "BIF" or 'banded iron formation' as agate due to the often well-developed alternating, red-and-gray colored banding. Banded iron formation is relatively common in glacial Superior Lobe glacial deposits but it doesn't have the overall translucent appearance. Sometimes specimens will have translucent chert layers between the red iron-rich bands. Even though they are not true Lakers, they can be quite beautiful and still worth carrying home.

Banded iron formation (BIF)

Jasper
This rock type is composed of nearly 100% crypto-crystalline quartz, but does not have the characteristic banding and generally forms in massive deposits as opposed to vesicle-filling of volcanic rocks. Jasper can be brightly-colored and pretty and some pieces can still be used for lapidary.

Banded jasper

Chert
A very common variety of crypto-crystalline quartz that usually occurs as a secondary replacement within sedimentary rocks. Color varies from colorless to white and yellow with some specimens exhibiting bright reds and oranges. Chert can also exhibit poorly defined banding which is a tip-off of an agate impersonator.

Chert

Silicified Limestone

This rock type is very common and often has a similar appearance to chert. Silicified limestone will often retain structures of the original material secondary quartz has replaced. Examples include sedimentary bedding — which is often confused with agate banding, and fossils such as coral. Sometimes limestone cavities are replaced with areas of banded agate that is usually white or colorless. These areas are often associated with irregular-shaped, tiny quartz crystal-lined open cavities called vug pockets. Even though they are not Lakers, some silicified limestone specimens can be quite beautiful.

Silicified limestone

Petrified Wood

Many people don't realize how common petrified wood is in glacial deposits. Pieces normally have dull white, yellow and brown colors and usually retain the evenly-spaced annual rings that are often mistaken for agate banding.

Petrified wood

"Leaverite"

Although the term has been jokingly used by rock-hounds for decades to mean "Leave er' right there," it is essentially what collectors should probably do with most material they pick up early on in their collecting careers. However, with a little persistence, it doesn't take long for a collector's eye to figure out what is agate and what is not!

Conclusion

We hope you have enjoyed seeing and reading about Lake Superior agates and that what you have learned will help with your collecting experience in this wonderful hobby. With even a minimal amount of effort, you will find agates and possibly even the highly coveted, big, beautiful "red and white."

Keep in mind, that no matter how much time you put in or how hard you look, the "all-timers" are highly elusive. Remember to enjoy yourself when searching and take in all the wonder and beauty of the outdoors and its splendor. Whether picking alone or with friends or family, the time spent searching is enriching and precious. It is at those moments when lost in thought and appreciating the natural world that you'll likely stumble upon that once-in-a-lifetime find!

Glossary

Agate – A waxy variety of cryptocrystalline quartz in which the colors occur in bands, clouds or distinct groups.

Amygdule – A gas cavity or vesicle in volcanic rocks, filled with secondary products such as zeolites, calcite, and silica minerals.

Basalt – **Dark volcanic** rock comprised of fine-grained minerals.

Bleached agate – Whitening or loss of color due to prolonged exposure to sunlight.

Calcite ($CaCO_3$) – One of the most common minerals; the principal constituent of limestone.

Chalcedony – Cryptocrystalline quartz, the mineral of agate and chert.

Composition – An aggregate, mixture, mass, or body formed by combining two or more substances; the chemical constituents of a rock or mineral; the mineralogical constitution of a rock.

Conchoidal fracture – A rock type or mineral fracture giving smoothly curved surfaces, characteristic of quartz.

Cross-section – A profile portraying an interpretation of a vertical section of the earth explored by geophysical or geological methods.

Crystal impression – A mold or cast left protruding into an agate, usually made by calcite before agate formation.

Deposition – The precipitation of mineral matter from solution as the deposition of agate or vein quartz.

Drift – Any rock material, such as boulders, till, gravel, sand or clay, transported by a glacier and deposited by or from the ice, or in water derived from the melting of the ice.

Erosion – The processes by which earthy or rock material is loosened or dissolved and removed from any part of the earth's surface.

Eye – The perfectly round, circular banded pattern found on some agates.

Face – A term applied to a well exposed banded area on an agate.

Fill hole – That area or areas where solutions entered and exited the vesicle..

Floating bands – Bands of chalcedony bound on both sides by clear crystal quartz.

Fortification – A term applied to pattern of agate banding often resembling the enclosing nature of a fort.

Fracture – The manner of breaking and appearance of a mineral when broken, which is distinctive for certain minerals.

Frost action – The weathering process caused by repeated cycles of freezing and thawing.

Gemstone – A general term for any precious or semiprecious stone.

Geode – A hollow, globular body, with an interior lining of inward-projecting crystals.

Glacier – A mass of ice with definite lateral limits, with motion in a definite direction and originating from the compacting of snow by pressure.

Holocene, Recent – that period of time since the last ice age.

Husk (slang) – A general term applied to the unabraded exterior of an agate.

Iron-oxide – Any mineral containing iron (Fe) and oxygen (O), including hematite (Fe_2O_3), magnetite (Fe_3O_4), goethite (FeOOH), and limonite.

Lapidary – A skilled work of cutting and polishing gems or other stones.

Laurentide ice sheet A mass of continental ice centered, during the Pleistocene Epoch, over what is now Hudson Bay.

Limonite – A mineral; field term for a group of brown, amorphous, naturally occurring, hydrous ferric minerals. Often occurs as a yellow, mustard-colored stain on agate husks.

Magma – A molten rock, formed within the crust or the upper mantle of the earth, which may consolidate to form an igneous rock.

Amazing Agates Glossary

Mineral – A structurally homogenous solid of definite chemical composition formed by the inorganic processes of nature.

Moss Agate – A variety of agate containing solid moss-like masses of manganese-oxides entombed by massive or banded chalcedony.

Oxidation – A process of combining with oxygen; removal of one of more electrons from an ion or an atom.

Painted agate – A heavily and stained agate that appears painted.

Paradise Beach agate – A type of agate originating from Paradise Beach location on the North Shore of Lake Superior, in Minnesota. Many contain small stringers of copper.

Peeled agate (slang) – An agate with fracturing between bands along the banding plane.

Permeable – Having a texture that permits water to move through it perceptively under the head differences ordinarily found in subsurface water.

Pleistocene – The earlier of two epochs comprising the Quaternary Period.

Precambrian – All rocks formed more than 600 million years ago.

Quaternary Period – The most recent time period, beginning 2 million years ago and continuing into the present.

Rifting event – The formation of a deeper fracture or break in the earth where magma upwellings occur together with spreading along the rift and the creation of new volcanic rock.

Sagenite – An acicular variety of rutile occurring in groups of crystals crossing at 60 degrees and often enclosed in quartz or other minerals.

Shadow agate – A type of agate that exhibits the optical effect created by the perception of depth between parallels bands of chalcedony.

Smoky quartz – A smoky, black- to brown-colored crystalline variety of quartz caused by exposure to natural radiation.

Solid agate (slang) – An agate banded completely with no clear quartz present.

Stain (slang) – Refers to red, orange and/or yellow colors due to oxidation of iron minerals.

Stratified – Formed or lying in beds, layers or strata.

Superior lobe – The lobe of ice that followed the depression or trough of Lake Superior into Minnesota, carrying Lake Superior agates.

Translucence – Admitting the passage of light, but not transparent.

Tube agate – A variety of agate with usually parallel, linear projections of mineral matter into the vesicle.

Tumble – To polish agates inside rotating drum with polishing grit.

Vesicle – A small cavity in an igneous rock, formed by the expansion of a bubble of gas or stream during the solidification of the rock.

Volcanic – Produced, influenced, or changed by a volcano or by volcanic processes.

Water-level agate – A variety with flat parallel banding topped with a fortification pattern or clear quartz.

Weathering – The group of processes, such as the chemical action of air, rainwater, plants and bacteria in addition to the mechanical action of changes of temperature, whereby rocks on exposure to the weather change in character, decay, and finally crumble into soil.

Wisconsin Glaciation – The last of the four classical glacial stages in the Pleistocene in North America.

Index

A
Agate Features
—banding 10
—copper 10
—crystal impressions 12
—fill holes 12
—iron oxide 11
—pinching 11
—shadows 11
—waterwashed 12
Agate types
—Australian 28
—Botswana 28
—Brazilian 28
—Chinese 29
—dryhead 28
—Fairburn 27
—Laguna 27
—Montana moss 27
—Thunder Bay 29
amorphous 2
amygdules 6
Anderson, George 24
Anderson, Jake 24
aragonite 12
arrowheads 21
Australian agates 28

B
banded iron formation 30
banded jasper 30
barite 12
basalt 4
Batman 22
beaches 16
Berghius, Lyle 23
BIF 30
Black Hills 27
blue agate 9
Botswana agates 28
Brazilian agates 28
brecciated agate 13

C
cabbing 17-18
cabochon 18
calcite 12
Calumet, Michigan 10
Canada 29
candy striper 13
Cash & Treasures 24
chalcedony 8, 10
chert 30
Chinese agates 29
clams 21
cleavage 21
columnar joints 4
conchoidal fracturing 21
copper 10
copper agate 10
Counsell, Ernie 22
Cox, Nina 22
crystal impressions 12

D–E
Doroff, Jim 25
dryhead agates 28
earthquakes 13
Ellsworth, Wisconsin 22
equipment 14
eye agate 8

F
face polishing 18
Fairburn agates 27
fill holes 12
Flaim, George 23
flint knapping 21
floater agate 7
Foothills State Forest 25
fortification agate 7
fur trade 21

G
Glacial Lake Duluth 5
glaciers 2, 3, 5, 6
goethite 9
gravel pits 15
Gum, Kirsten 24

H
hand rake 14
Harris, John 26
Howard, Matt 25
Hudson Bay 5

I
Interstate State Park 3
Iowa 6
iron oxide 10, 11
Island Lake, Minnesota 23

J–K
jasper 30
Kansas 3, 6
Kirkorkian, Dave 25
knapping 21
Kreminski, Peter 23

L
Laguna agates 27
Lake Nipigon 3
Lake Superior agates
—Batman 22
—Berghius 23
—blue 9
—Brush Fire 26
—Counsell 22
—Doroff 25
—Engagement 23
—eye 8
—Famous agates 22-26
—Features of 10-13
—floater 7
—fortification 7
—Ham 22
—Home Run 26
—How they form 3
—How to display 17

Amazing Agates Index

—How to value 19-20
—Howard 25
—Howling Wolf 25
—Look alikes 30-31
—Minnehaha Falls 23
—mosaic 9
—moss 8
—painted 7
—paintstone 7
—Peanut Butter & Jelly 23
—peeler 9
—Perfect Pounder 26
—Red & White 24
—Reef Agates 24
—Remember Me? 24
—Rolling Stone 24
—sagenite 9
—skip-an-atom 9
—tube 8
—Types 7-9
—water-level 8
—where to find 15-16
Lakeland, Minnesota 26
lakers 2
lapidary services 18
lava 3, 4
"leaverite" 31
loupe 14

M

magnifying glass 14
Mammoth 21
marbles 18
Mastadon 21

Mexico 27
Michigan 10
Midcontinent Rift 3, 5
Milaca, Minnesota 23
mineral oil 17
Minnehaha Falls 23
Minnesota
—Island Lake 23
—Lakeland 26
—Milaca 23
—Pierz 22
—Randall 24
—Saginaw 26
—Scanlon 25
—Swanville 26
—Two Harbors 23
Minnesota's State Gemstone 1
Montana 28
Montana moss agates 27
Moose Lake Agate Days 24, 26
Morell, John 24
mosaic agate 9
moss agate 8

N–O

Nebraska 6
New Haven, Iowa 22
obsidian 21
Ontario 29
onyx 8

P

painted agate 7
paintstone 7

Paleo Indians 21
peeler agate 9
"petrified ham" 22
petrified wood 31
Pierz, Minnesota 22
pinching 11
polishing 18
Post, Cody 22
Precambrian Era 3

Q–R

quartz 2
Randall, Minnesota 24
rhyolite 6, 27
rift zone 3
rifting 3
rock tumblers 17
Rodeski, Jim 26
Rolling Stones 24
ruin agate 13
rutile 9, 12

S

sagenite agate 9
Saginaw, Minnesota 26
scalpels 21
Scanlon, Minnesota 25
Schussler, Bobby 26
shadow effect 11
silica 5
silicified limestone 21, 31
skip-an-atom agate 9
slab cutting 18
South Dakota 27

spear heads 21
Stone Age 21
storm ridge 16
storm terraces 16
Superior Lobe 5
Swanville, Minnesota 26

T

Thunder Bay agates 29
tube agate 8
tumbling 17
Twardzik, George 24
Two Harbors, Minnesota 23

U–V

Upper Peninsula, Michigan 10
vesicles 3, 4
volcanism 13

W–Z

water-level agate 8
waterwashed agate 12
Wisconsin 6
Wisconsin Glaciation 5
Wolter, Janet 23
Yangtze River 29
zeolites 12

Titles of Interest

Daniel, Glenda, and Sullivan, Jerry, *A Sierra Club Naturalist's Guide: The North Woods*. San Francisco, CA: Sierra Club Books, 1981.

Green, John C., *Geology on Display: Geology and Scenery of Minnesota's North Shore State Parks*. St. Paul, MN: Minnesota Department of Natural Resources, 1996.

LaBerge, Gene L., *Geology of the Lake Superior Region*. Tucson, AZ: Geoscience Press, 1994.

Lynch, Bob. *Lake Superior Rocks & Minerals: A Field Guide to the Lake Superior Area*. Cambridge, MN: Adventure Publications, 2008.

Ojakangas, Richard W., *Roadside Geology of Minnesota*. Mountain Press Publishing Company, Missoula MT, 2009

Ojakangas, Richard W. and Matsch, Charles L., *Minnesota's Geology*. Minneapolis, MN: University of Minnesota Press, 1982.

Pough, Frederick H., *Peterson Field Guide: Rocks and Minerals*. Boston, MA: Houghton-Mifflin Company, 1988.

Roberts, David C., *Peterson Field Guide: Geology Eastern North America*. Boston, MA: Houghton-Mifflin Company, 1996.

Sansome, Constance Jefferson, *Minnesota Underfoot*. Stillwater, MN: Voyageur Press, 1983.

Sorrell, Charles A., *Minerals of the World: A Golden Guide*. New York, NY: Golden Press, 1973.

Stensaas, Sparky, *Rock Picker's Guide to Lake Superior's North Shore: Second Edition*. Duluth, MN: Kollath-Stensaas Publishing, 2000.

Other Books by Scott Wolter

Wolter, Scott F., *The Hooked X: Key to the Secret History of North America*, North Star Press of St. Cloud, Inc., St. Cloud, MN, 2009

Wolter, Scott F., *The Lake Superior Agate: Fourth Edition*, Lake Superior Agate Inc., Minneapolis, MN, 2008, 2001, 1996, 1994, 1986.

Wolter, Scott F., *The Kensington Rune Stone: Compelling New Evidence*, Lake Superior Agate Publishing, Chanhassen, MN, 2006.

Wolter, Scott F., *Structural Condition Assessment*, Chapter 16: Concrete, John Wiley & Sons, Inc., Hoboken, NJ, 2005.

Wolter, Scott F., *The Lake Superior Agate: One Man's Journey*, Outernet Publishing LLC, Eden Prairie, MN, 2001.

Wolter, Scott F., *Ettringite: Cancer of Concrete*, American Petrographic Services Inc., St. Paul, MN 1997.

OTHER GREAT KOLLATH+STENSAAS BOOKS

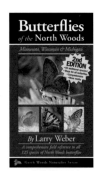

Butterflies of the North Woods: 2nd Edition *by Larry Weber* Revised and more comprehensive than ever, this is your guide to all 120 species of butterflies found in the North Woods of Minnesota, Wisconsin and Michigan. 260 color photos. 288 pages, softcover, $18.95

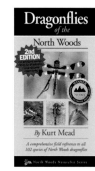

Dragonflies of the North Woods *by Kurt Mead* Winner of the National Outdoor Book Award, this newly revamped guide will lead you to all 103 species of North Woods dragonflies. It's the first guide exclusively for MN, WI and MI. 200-plus color photos, 193 pages, softcover, $18.95

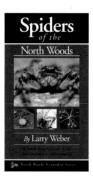

Spiders of the North Woods *by Larry Weber* Meet over 60 fascinating northern spiders. Spider guides are few and far between and this is one of the best. Larry's easy-to-follow format makes field identification simple and fun. 212 color photos, 95 illustrations, 205 pages/softcover, $18.95

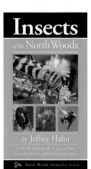

Insects of the North Woods *by Jeffrey Hahn* Insects of the North Identify 444 species of six-legged critters native to MN, WI and MI. 600-plus color photos of beetles, bees, wasps, grasshoppers, crickets, true bugs, moths, flies, butterflies, dragonflies and mayflies. 245 pages/softcover, $18.95

Orchids of the North Woods *by Kim and Cindy Risen* A guide to all 50 species of orchids that survive and thrive in the North Woods. Includes exact range maps for MN, WI and MI, public locations to search, close-up photos of the flowers and many photos of seed pods. 150-plus color photos, 140 pages/softcover, $18.95

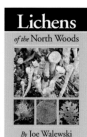

Lichens of the North Woods *by Joe Walewski* Lichens are the unsung super heros of the Northern wilds; they are tiny organisms that have the capability to dissolve solid granite. One hundred eleven species are shown in 150-plus beautiful color photos. 152 pages/softcover, $18.95

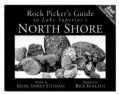

Rock Picker's Guide to Lake Superior's North Shore *by Mark Sparky Stensaas* Over 60,000 sold! This popular and easy-to-use guide helps rock-enthusiasts of every age identify all those rocks seen on the shores of Lake Superior. 12 detailed maps show the best beaches for picking. Includes illustrations of each type of rock, a description, tips for recognizing it and where to go to find it. 65 illustrations, 43 pages/softcover, $9.95

Fascinating Fungi of the North Woods *by Cora Mollen & Larry Weber* This is the first guide exclusively for MN, WI & MI. All 118 species in this book are represented in beautifully detailed, full-color illustrations. Spore-color icons aid in identification; phenograms show the season you can expect to see them. Over 200 color illustrations. 113 pages/softcover, $14.95

Find out more about these books and all our other titles at **www.kollathstensaas.com**